Inspire Science

Be a Scientist Notebook

Grade K

Mc
Graw
Hill
Education

ConnectED.mcgraw-hill.com

Send all inquiries to:
McGraw-Hill Education
8787 Orion Place
Columbus, OH 43240

ISBN: 978-0-02-135693-5
MHID: 0-02-135693-9

Printed in the United States of America.

1 2 3 4 5 6 7 8 9 QVS 19 18 17 16 15

You don't have to be a grown-up to be a scientist. You can think and act like a scientist every day. Scientists think and act certain ways. We call thinking and acting like a scientist or engineer a practice. You will use these practices in your **Be a Scientist Notebook!**

Emily

Ask Questions

Emily is full of questions about how things fly. She took her first plane ride to visit her grandparents. She wrote down all the questions she had about the plane. Asking questions is important for scientists. It helps them make a plan to answer them. You will ask many questions in your *Be a Scientist Notebook.*

Kayla

Use a Model

Kayla loves gardening. She draws out a map of what she wants in her garden before she plants the vegetables. She gets an idea of what things should look like and where she wants to plant the vegetables. She is using a model. You will make and use models to test ideas in your *Be a Scientist Notebook.*

Carry Out an Investigation

Deven

Deven likes to make a lot of noise. He loves the sounds he hears in movies. He tries to recreate some of the sounds with things he has in his house. When he tests each item to see what it sounds like, he is carrying out an investigation. He can use what he learns to make changes and get better sounds. You will carry out investigations in your *Be a Scientist Notebook.*

Interpret Data

Marisol

Marisol wants to be a paramedic. She will need to use what she knows about her patients to determine what is wrong with them. This is called interpreting data. You will need to look at information you gather and interpret that data in your *Be a Scientist Notebook.*

Use Math

Chloe

Chloe is a builder. She wants to be a carpenter when she grows up. When building something, measuring is important. Sometimes she needs to use math to make sure the measurements are just right. She always measures everything twice just to make sure. You will use math in your *Be a Scientist Notebook.*

Design a Solution

Erik

Erik sometimes has trouble learning in school. At home, he loves to play games. He came up with an idea to help him learn using a game. He solved a problem by designing a solution. He wants to do this as a video game designer when he grows up. You will design solutions to help solve problems in your *Be a Scientist Notebook.*

Hugo watches the sky every day. He has noticed that certain kinds of clouds lead to certain kinds of weather. When he sees the clouds, he waits to see what kind of weather occurs. The type of weather that occurs is evidence that supports his thoughts. You will use evidence to support your thoughts in the **_Be a Scientist Notebook._**

Hugo

Gather Information

Jordan has many pets and he loves to train them to do tricks. He uses the Internet and books from the library to learn the best ways to train them. He wants to be an animal trainer. He is gathering information to learn more about something. You will gather information in your **_Be a Scientist Notebook._**

Jordan

The Career Kids love science and want to have important jobs in science and engineering someday. In _Inspire Science,_ they will help you learn about many different science careers. You can have an important science career someday, too. But you can start thinking like a scientist or an engineer today!

Table of Contents

Science in My World
Forces and Motion

What do you wonder about
the ramp?

_____

\- \-

_____

\- \-

_____

\- \-

_____

Look and listen for these words as you learn
about forces and motion.

Words in Science

push	pull	position
motion	speed	distance
collide	energy	force

Name _____ Date _____

How do ramps help you move things?

Chloe Carpenter

Show how you think ramps make objects move. Draw a picture below.

Science in Action

I will do an experiment.
I will collect data.
I will see cause and effect.

Name _____ Date _____

Science in My World
Pushes and Pulls

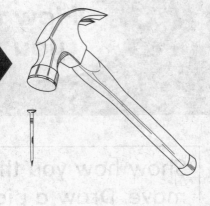

The hammer hits the nail. What do you wonder about the hammer and nail?

_____
- -
_____
_____
- -
_____
_____
- -
_____
_____
- -
_____
_____
- -
_____

Essential Question

 How are pushes and pulls different from each other?

Chloe Carpenter

4   Lesson 1 Pushes and Pulls

What is a push? What is a pull?
Draw or write what you think.

Science in Action

I will do an experiment.
I will collect data.
I will see cause and effect.

Name _____ Date _____

Inquiry: Push and Pull

What happens when you push harder?
What happens when you pull harder?

Think about the activity. Fill in the cause
and effect chart.

Cause	Effect
1. The monkey pulled the wagon handle. He pulled it hard.	1. The dog in the wagon went over the finish line!
2.	2.
3.	3.

Name _____ Date _____

Represent Draw and label a push.

┌───┐
│   │
│   │
│   │
│   │
│   │
│   │
│   │
└───┘

Represent Draw and label a pull.

┌───┐
│   │
│   │
│   │
│   │
│   │
│   │
│   │
└───┘

Name _____ Date _____

Complete the "I can..." statement.

I can do an
experiment

I can collect
data

I can see cause
and effect

Name _____ Date _____

How do you open a car door?

Express Draw and label your answer.

Name _____ Date _____

Look at the picture. Look at the words.
Circle the word that goes with the picture.

1.

pull   push

2.

pull   push

3.

pull   push

4.

pull   push

Name _____ Date _____

Think about pushes and pulls. Complete the chart.

Cause		Effect
1. I pull on the refrigerator door handle.	→	1. The refrigerator door opens.
2.		2.
3.		3.

Name _____ Date _____

Essential Question

How are pushes and pulls different from each other?

_____

- -

_____

- -

_____

- -

_____

- -

_____

- -

_____

- -

_____

Word Box

push   pull   position   motion

Name _____ Date _____

Think about the hammer hitting the nail.

What is a push? What is a pull?
How are they different?

Draw or write your answer.
Use the words from the word box.

Science in Action

Rate Yourself. Circle the image that tells how well you did.

I did do an experiment.

 Almost Always   Sometimes   Still Learning

I did collect data.

 Almost Always   Sometimes   Still Learning

I did see cause and effect.

 Almost Always    Sometimes   Still Learning

Name _____ Date _____

Science in My World
Strength and Distance

Think about the moms pushing their children. What do you wonder about this action?

_____
- -
_____
- -
_____
- -
_____
- -
_____

Essential Question

What happens if you push or pull an object harder?

Chloe Carpenter

What makes you go high on a swing? Draw or write what you think.

Science in Action

I will do an experiment.
I will collect data.
I will see cause and effect.

Name _____ Date _____

Inquiry: Slow Down or Speed Up

Find a classroom object to test. Pick the material for it to slide on. Test it. Collect data from your experiment. Record it below.

	Material 1: _____	Material 2: _____
First try		
Second try		

Name _____ Date _____

Interpret Look around your classroom.
Draw and label something you can push.
Draw and label something you can pull.

Name _____ Date _____

Complete the "I can…" statements.

I can do an experiment

I can collect data

I can see cause and effect

Name _____ Date _____

Justify A mom is pushing her daughter on a swing.

1. The mom gives her daughter a soft push. What happens?

_____

- -

_____

2. The mom gives her daughter a harder push. What happens?

_____

- -

_____

Name _____ Date _____

Inquiry: How Hard to Push

You need to get your object over the line.
How hard do you have to push?
Make a prediction.

1. Circle the object you will test.

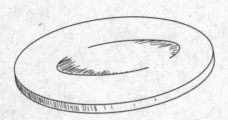  

2. You will push your object to the first line.
   How hard will you have to push?

   hard   harder   hardest

3. You will push your object to the second
   line. How hard will you have to push?

   hard   harder   hardest

Name _____ Date _____

Inquiry: How Hard to Push

What happens when you push the object?
Record what you see in the boxes below.

	Object: _____
First Line	
Second Line	

Name _____ Date _____

Essential Question

What happens if you push or pull an object harder?

_____
- -
_____
- -
_____
- -
_____
- -
_____
- -
_____
- -
_____

Word Box

speed   distance

Name _____ Date _____

Think about the mom with her daughter on the swing.

How will the girl go higher on the swing?
Draw or write your answer.
Use the words from the word box.

Science in Action

Rate Yourself. Circle the image that tells how well you did.

I did do an experiment.

 Almost Always   Sometimes   Still Learning

I did collect data.

 Almost Always   Sometimes   Still Learning

I did see cause and effect.

 Almost Always   Sometimes   Still Learning

Name _____ Date _____

Science in My World
When Objects Collide

What do you wonder about how the ball moves when it is kicked?

_____

- -

_____

- -

_____

_____

- -

_____

_____

- -

_____

Essential Question

What happens when objects touch or collide?

Chloe Carpenter

24   Lesson 3 When Objects Collide

Name _____ Date _____

What happens when two things bump into each other? Draw or write what you think.

Science in Action

I will do an experiment.
I will collect data.
I will see cause and effect.

Name _____ Date _____

Inquiry: Marbles Collide

What happens when two marbles bump into each other? Record what you saw below.

	2 marbles the same size	1 big marble and 1 small marble
Roll 1		
Roll 2		

Name _____ Date _____

When Objects Collide

Draw a picture of objects colliding. Write a sentence to explain what is happening.

Name _____ Date _____

Complete the "I can..." statements.

I can do an experiment

I can collect data

I can see cause and effect

Name _____ Date _____

Cause and Effect of Collisions

Justify Think about collisions. Fill in the cause and effect chart.

Cause	Effect
1. I pushed the bowling ball into the pins.	1. The bowling pins were knocked down.
2.	2.
3.	3.

Name _____ Date _____

Inquiry: Balls Colliding

What happens when balls collide?
Draw your prediction.

Name _____ Date _____

Inquiry: Balls Colliding

What happens when balls collide?
Collect data from your experiment.
Record it below.

	Ball 1: _____	Ball 2: _____
Roll 1		
Roll 2		

	Ball 1: _____	Ball 2: _____
Roll 1		
Roll 2		

Name _____ Date _____

Essential Question

What happens objects touch or collide?

_____

- -

_____

- -

_____

- -

_____

- -

_____

- -

_____

- -

_____

- -

_____

Word Box

collide

Name _____ Date _____

Think about the foot kicking the soccer ball.

What happens when things collide in sports? Draw or write your answer. Use the words from the word box.

Science in Action

I did do an experiment.

Rate Yourself. Circle the image that tells how well you did.

 Almost Always   Sometimes   Still Learning

I did collect data.

 Almost Always   Sometimes   Still Learning

I did see cause and effect.

 Almost Always   Sometimes   Still Learning

Name _____ Date _____

Science in My World
Direction and Forces

Write a question about the dominoes.

_____

- -

_____

- -

_____

_____

- -

_____

_____

- -

_____

_____

- -

_____

Essential Question

How can pushes and pulls change an object's direction?

Chloe Carpenter

Name _____ Date _____

Do you think you can make the dominoes fall over? Draw or write what you think.

Science in Action

I will do an experiment.
I will collect data.
I will see cause and effect.

Name _____ Date _____

Inquiry: All Fall Down

How can you make 10 dominoes fall down?
Draw what your pattern will look like.
Label any materials you will use.

Name _____ Date _____

Inquiry: All Fall Down

Test your domino pattern.
Record what you saw below.

Name _____ Date _____

Complete the "I can..." statements.

I can do an experiment

I can collect data

I can see cause and effect

Name _____ Date _____

Represent Draw two examples of objects that can be pushed, pulled, and change directions. Pick objects you see every day. Label your drawing.

Name _____ Date _____

Inquiry: Dominoes Changing Direction

How can you make 10 dominoes change direction? Draw what your pattern will look like. Label any materials you will use.

Name _____ Date _____

Inquiry: Dominoes Changing Direction

Test your domino pattern. Record what you saw below.

Name _____ Date _____

Essential Question

How can pushes and pulls change an object's direction?

_____

- -

_____

- -

_____

- -

_____

- -

_____

- -

_____

- -

_____

Word Box

energy   force

Name _____ Date _____

Think about the dominoes falling over.

What makes the dominoes fall over?
What makes the dominoes change
direction? Draw or write your answer.
Use the words from the word box.

Science in Action

Rate Yourself. Circle the image that tells how well you did.

I did do an
experiment.

Almost   Sometimes   Still
Always   Learning

I did collect data.

Almost   Sometimes   Still
Always   Learning

I did see cause
and effect.

Almost   Sometimes   Still
Always   Learning

Name _____ Date _____

How do ramps help you move things?

_____

- -

_____

- -

_____

- -

_____

- -

_____

- -

_____

- -

_____

- -

_____

- -

_____

It's a Wrap!
Forces and Motion

Design a Ramp

Design a ramp that will help someone in a wheelchair go upstairs. Label the materials you will need to make your ramp.

Forces move everything around you!

Science in My World
Energy and the Sun

What do you wonder about the playground?

_____

- -

_____

- -

_____

- -

_____

- -

_____

- -

_____

Words in Science

Sun   Earth   warm   cool   shade

How does the Sun affect the playground?

Hugo
Meteorologist

Think about your playground. Draw what you saw and did.

Science in Action

I will collect data.
I will make comparisons.
I will design a solution.
I will see cause and effect.

Name _____ Date _____

Science in My World
Sunlight and Earth's Surface

What do you wonder about the Sun?

_____

- -

_____

- -

_____

- -

_____

- -

_____

Essential Question
What does the Sun do for Earth?

Hugo
Meteorologist

Name _____ Date _____

What do you think the Sun does for you?
Draw or write what you think.

Science in Action

I will collect data.
I will make comparisons.
I will see cause and effect.

Name _____ Date _____

Inquiry: Sunlight and Water
Does sunlight make water warmer?

	Water	**Water in Sunlight**
Start		
After 10 minutes		

Name _____ Date _____

Connections in Science

Cause and Effect What effect did the sunlight have on the water?

_____

- -
_____
_____
- -
_____
_____
- -
_____
_____
- -
_____
_____
- -

Talk About It

What do you think would happen to the water if it was left in the sunlight all day?

Name _____ Date _____

Think about all of the ways the Sun affects Earth. Write one thing the Sun does in each circle.

What does the Sun do?

Name _____ Date _____

Complete the "I can..." statement.

I can <u>collect data</u>

I can <u>compare data</u>

I can <u>see cause</u>
<u>and effect</u>

Name _____ Date _____

Talk About It

Think of a time when you were in the sunlight. What did it feel like?

Do you think the Sun will warm sand, soil, and rocks? Circle yes or no.

Sand	Yes	No

Soil	Yes	No

Rocks	Yes	No

Name _____ Date _____

Inquiry: Sunlight and Earth's Surface

Touch the objects before and after being in sunlight. How do they feel?

	Before Sunlight	After Sunlight
Sand		
Soil		
Rocks		

Name _____ Date _____

Essential Question
What does the Sun do for Earth?

_____
- -
_____
- -
_____
- -
_____
- -
_____
- -
_____

Word Box

Sun   Earth   warm   cool

Name _____ Date _____

Think about Hugo's question.

What does the Sun do for you?
Draw or write your answer.
Use the words from the word box.

Science in Action

I did collect data.

Rate Yourself. Circle the image that tells how well you did.

Almost Always   Sometimes   Still Learning

I did make comparisons.

Almost Always   Sometimes   Still Learning

I did see cause and effect.

Almost Always   Sometimes   Still Learning

Lesson 1 Sunlight and Earth's Surface   **57**

Name _____ Date _____

Science in My World
Sunlight and Shade

What do you wonder about the umbrellas on the beach?

_____

- -

_____

_____

- -

_____

- -

_____

- -

_____

Essential Question

What could help you stay cool in the Sun?

Hugo
Meteorologist

Name _____ Date _____

Hugo was playing outside. The Sun was hot.
Hugo is hot. What should Hugo do? Draw
or write what you think.

Science in Action

I will design a solution.
I will see cause and effect.

Name _____ Date _____

Inquiry: Temperatures Throughout the Day
Record your choices from the activity.

Time of Day	+	Shade	=	Temperature

 +  =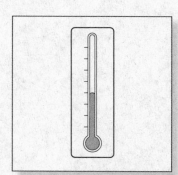

+  =

+  =

Name _____ Date _____

Connections in Science

Cause and Effect How did shade affect the temperature?

_____

- -

_____

_____

- -

_____

_____

- -

_____

Talk About It

What are some things outside that make shade?

Lesson 2 Sunlight and Shade   61

Name _____ Date _____

Complete the "I can..." statement.

I can design a solution

I can see cause and effect

Name _____ Date _____

Animals in the Shade

1. Do animals need shelter from the Sun?
   Draw or write your answer.

```



```

2. Draw or write an example of how animals
   protect themselves from the Sun.

```



```

Name _____ Date _____

Animal Shelter
Choose a camel or lizard. Circle your choice.

Draw the materials you will need for your animal's shelter. Label each material.

Name _____ Date _____

Animal Shelter

Draw and label your shelter.

Talk About It

What problem will your shelter solve?

Name _____ Date _____

Essential Question

What could help you stay cool in the Sun?

_____

- -

_____

- -

_____

- -

_____

- -

_____

- -

_____

Word Box

shade

Name _____ Date _____

Think about Hugo in the Sun.

What could help keep him cool? Draw or write what you think. Use the word from the word box.

Science in Action

I did design a solution.

Circle the image that tells how well you did.

Almost Always   Sometimes   Still Learning

I did see cause and effect.

Almost Always   Sometimes   Still Learning

Name _____ Date _____

Talk About It

How does the Sun affect the playground?

Draw your playground. Draw a structure that could make shade to keep you cool while you play.

It's a Wrap!
Energy and the Sun

Name _____ Date _____

Open Inquiry in Science

Shade helps keep us cool. Look at your back yard. Look for a place that needs shade. How would you build something to create shade? Draw what you would build below.

The Sun warms the Earth and all living things!

Science in My World
Weather

There is lightning in the sky.
What do you wonder about the
lightning?

_____

- -

_____

_____

_____

- -

_____

- -

_____

Look and listen for these words as you learn
about weather.

Words in Science

weather	season	thermometer
temperature	forecast	severe weather

How does the weather affect people?

Hugo
Meteorologist

Show how you think weather affects you.
Draw a picture below.

Science in Action

I will get information.
I will look at data.
I will see patterns.

Name _____ Date _____

Science in My World
Weather Patterns

Winter

Summer

The tree changes in each season. What do you wonder about the tree?

_____

- -

_____

- -

_____

- -

_____

- -

_____

Essential Question

What words can you use to describe weather?

Hugo
Meteorologist

Name _____ Date _____

What is weather? What are weather patterns? Draw or write what you think.

Science in Action

I will get information.
I will look at data.
I will see patterns.

Name _____ Date _____

Inquiry: Wind Effects

Draw or write your prediction. Then, draw or write what you saw.

	Object 1: _____	Object 2: _____
Prediction		
What you saw		

Name _____ Date _____

Inquiry: Weather Chart

Record the weather each day in the table below.

	S	M	T	W	Th	F	Sa
sunny							
windy							
rainy							
snowy							

Talk About It

What patterns do you see in your weather chart?

Name _____ Date _____

Complete the "I can…" statement.

I can get
information

I can look at data

I can see patterns

Name _____ Date _____

Everyday Weather

Represent Think of weather you see often.
Draw two kinds of weather. Label your
drawings.

Name _____ Date _____

Favorite Season

Represent What is your favorite season? Ask your classmates. Use tally marks. Fill in the table.

Fall	
Winter	
Spring	
Summer	

Name _____ Date _____

Nature Walk

Go on a nature walk outside. Look for three things that tell what season it is. Draw a picture of each thing that you saw.

1.

2.

3.

Name _____ Date _____

Essential Question

What words can you use to describe weather?

_____

- -

_____

_____

- -

_____

_____

- -

_____

_____

- -

_____

_____

- -

Word Box

weather	windy	rainy	cloudy
snowy	sunny	season	winter
summer	spring	fall	patterns

Name _____ Date _____

Think about the tree in the seasons.

What is weather? What are weather patterns? Draw or write what you think. Use the word from the word box.

Winter

Summer

Science in Action

Rate Yourself. Circle the image that tells how well you did.

I did get information.

Almost Always

Sometimes

Still Learning

I did look at data.

Almost Always

Sometimes

 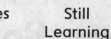
Still Learning

I did see patterns.

Almost Always

Sometimes

Still Learning

Name _____ Date _____

Science in My World
Describe Weather

Look at the picture. What do you wonder about this weather tool?

_____
_____
_____
_____
_____
_____
_____
_____
_____

Essential Question

How do tools help to describe the weather?

*Hugo
Meteorologist*

Name _____ Date _____

Make a list of tools we use to tell about the weather. Draw or write your answer.

Science in Action

I will get information.
I will look at data.
I will see patterns.

Name _____ Date _____

Inquiry: Weather Tool Matching

Weather tools measure kinds of weather.
Match the weather tool with the weather
it measures.

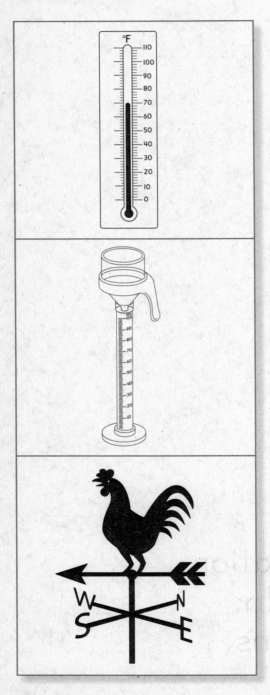

thermometer   rainfall

rain gauge   wind direction

wind vane   temperature

Name _____ Date _____

Inquiry: Measure the Temperature

Justify You are going to measure the temperature. You will measure it inside and outside. Which temperature will be warmer? Which temperature will be colder? Make your prediction below.

Inside Temperature	Outside Temperature

Talk About It

What do you know about temperature that makes you think this?

Name _____ Date _____

Inquiry: Measure the Temperature

Justify Record the temperatures below.
Circle the warmer temperature.

Inside Temperature	Outside Temperature

Talk About It

Would your results be different during a different season?

Name _____ Date _____

Complete the "I can…" statement.

I can get
information

I can look at data

I can see patterns

Name _____ Date _____

Measuring Objects

Interpret Measure two objects with a ruler.
Record your data below.

	Object 1: _____	Object 2: _____
Measurements		

How is measuring objects like measuring snow? How is measuring objects like measuring rain?

Name _____ Date _____

Number the Steps

Put the steps of measuring and comparing temperature in order.

Compare the temperature.

Put a thermometer outside.

Record the temperature each day.

Name _____ Date _____

How do tools help to describe the weather?

_____

- -

_____

- -

_____

- -

_____

- -

_____

- -

_____

- -

_____

- -

Word Box

thermometer   temperature

Name _____ Date _____

Think about the weather tools you have learned about.

How does a weather vane or another tool help us describe the weather? Draw or write what you think. Use the word from the word box.

Science in Action

Rate Yourself. Circle the image that tells how well you did.

I did get information.

Almost Always   Sometimes   Still Learning

I did look at data.

Almost Always   Sometimes   Still Learning

I did see patterns.

Almost Always   Sometimes   Still Learning

Name _____ Date _____

Science in My World
Forecast Weather

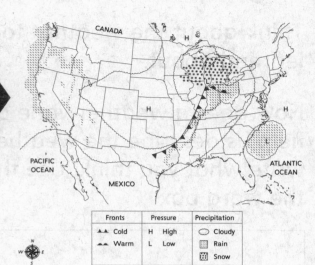

This is a weather map.
What do you wonder about
the picture?

_____
- -
_____
- -
_____
- -
_____
- -

Essential Question

How do people tell
what the weather
is going to be like?

Hugo
Meteorologist

Name _____ Date _____

Hugo is having a picnic on Saturday.
He is wondering if it will rain. How can
he find out? Draw or write your answer.

Science in Action

I will get information.
I will look at data.
I will see patterns.

Name _____ Date _____

Inquiry: Comparing Forecast to Actual Weather

Look at today's weather forecast. Look outside. Think about what is the same. Think about what is different. Fill in the chart.

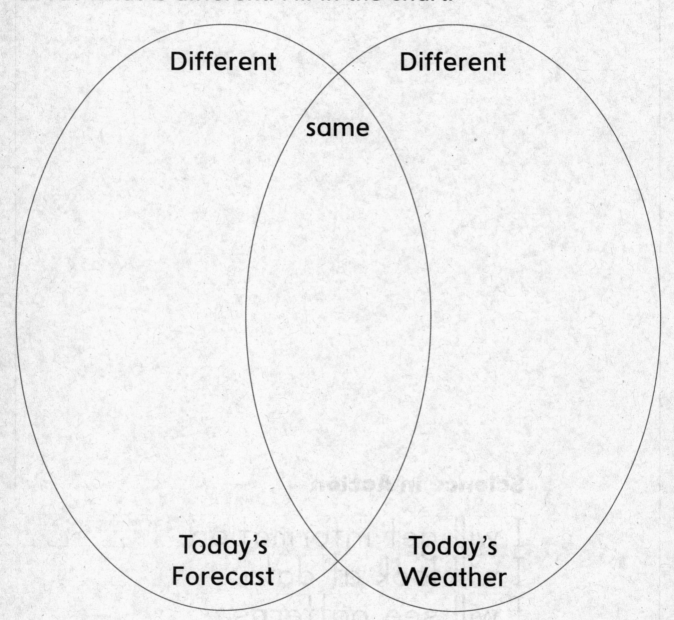

Different   Different

same

Today's   Today's
Forecast   Weather

Name _____ Date _____

Complete the "I can..." statement.

I can get
information

I can look at data

I can see patterns

Name _____ Date _____

Temperature Map

Interpret Think about the weather map your teacher showed. Pick a state. Fill in the chart below. Circle the higher temperature.

	State: _____	State: _____
Temperature		

Talk About It

What can you learn from comparing the temperatures?

Name _____ Date _____

Radar Map

Interpret Think about the radar map your teacher showed. Fill in the chart below.

Color	Meaning of Color
Green	
Yellow	
Orange	
Red	
Pink	
Blue	

Name _____ Date _____

Weather in my Area

What is weather like near me? Record your research below.

Name _____ Date _____

Write one fact about your favorite kind of weather. Draw your favorite weather. Draw one tool used to tell about your weather. Label your drawings.

_____
- -
_____
_____
- -
_____
_____
- -
_____

Name _____ Date _____

Essential Question

How do people tell what the weather is going to be like?

_____
- -
_____
- -
_____
- -
_____
- -
_____
- -
_____
- -

Word Box

forecast

Name _____ Date _____

Think about the weather map.

How do meteorologists tell what the weather is going to be like? Can you explain? Draw or write what you think. Use the word from the word box.

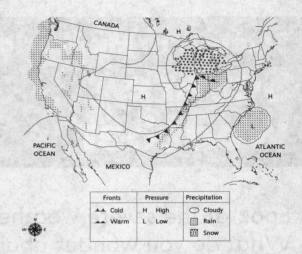

Science in Action

I did get information.

I did look at data.

I did see patterns.

Rate Yourself. Circle the image that tells how well you did.

Almost Always | Sometimes | Still Learning

Almost Always | Sometimes | Still Learning

Almost Always | Sometimes | Still Learning

Name _____ Date _____

Science in My World
Severe Weather

Look at the severe weather.
What do you wonder about it?

_____
- -
_____
_____
- -
_____
_____
- -
_____
_____
- -

Essential Question

What are some kinds of severe weather?

Hugo
Meteorologist

Name _____ Date _____

What are some kinds of bad weather?
How can we stay safe? Draw or write what
you think.

Science in Action

I will get information.
I will look at data.
I will see patterns.

Name _____ Date _____

Inquiry: Make Thunder

Talk About It

What happened when you hit the bag? What did the sound remind you of?

What is bad weather?

Think about a time when the weather was bad. Draw a picture of it. Label your picture.

Name _____ Date _____

Complete the "I can…" statement.

I can get information

I can look at data

I can see patterns

Name _____ Date _____

Tools for Severe Weather

Justify Circle the items you think will be useful in bad weather. Explain why.

flashlight	batteries	soccer ball
canned food	nail polish	water bottle

_____

- -

_____

- -

_____

- -

_____

Name _____ Date _____

A House in Severe Weather

Draw the weather in the scene to match each caption.

It is raining hard outside.

Name _____ Date _____

A House in Severe Weather

Draw the weather in the scene to match each caption.

There is a tornado!

Name _____ Date _____

A House in Severe Weather

Pick a different kind of severe weather. Draw the weather. Write your own caption.

_____

- -

_____

Name _____ Date _____

Essential Question

? What are some kinds of severe weather?

_____

- -

_____

_____

- -

_____

_____

- -

_____

_____

- -

_____

_____

- -

_____

Word Box

severe weather   tornado   thunderstorm
blizzard   hurricane

Name _____ Date _____

Think about severe weather.

What are some kinds of severe weather? How can you stay safe in severe weather? Draw or write what you think. Use the word from the word box.

Science in Action

Rate Yourself. Circle the image that tells how well you did.

I did get information.

 Almost Always   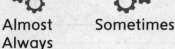 Sometimes   Still Learning

I did look at data.

 Almost Always   Sometimes   Still Learning

I did see patterns.

 Almost Always   Sometimes   Still Learning

Talk About It

How do we know what the weather will be like?

\- -

_____

\- -

_____

\- -

_____

\- -

_____

\- -

_____

\- -

It's a Wrap!
Weather

Name _____ Date _____

Let's Make a Forecast

Make a weather map and forecast for where you live.

Weather affects everything we do!

Science in My World
Plants and Animals

What questions do you have about plants and animals?

_____

_____

_____

_____

_____

_____

_____

_____

Look and listen for these words as you learn about protecting our Earth.

Words in Science

living	nonliving	need	soil
nutrients	survive	shelter	habitat
ecosystem			

What do plants and
animals need to live?

Jordan
Animal Trainer

Show what plants and animals need to live.
Draw and write what you think.

Science in Action

I will use models.
I will observe patterns.
I will know things in nature
work together.

Name _____ Date _____

Science in My World
Plant and Animal Needs

What questions do you have about what plants and animals need? Write a question.

_____

_____

_____

_____

_____

_____

 **Essential Question**
What do plants and animals need to live?

Jordan

Name _____ Date _____

How would you take care of a plant or animal? Draw and write what you think.

Science in Action

I will use models.
I will observe patterns.
I will know things in nature
work together.

Name _____ Date _____

Tell what you know. Circle the correct answer.

1. Does a plant or animal need water?

   plant   animal   both

2. Does a plant or animal need shelter?

   plant   animal   both

3. Does a plant or animal need light?

   plant   animal   both

4. Does a plant or animal need food?

   plant   animal   both

5. Does a plant or animal need air?

   plant   animal   both

6. Does a plant or animal need soil?

   plant   animal   both

7. Does a plant or animal need space?

   plant   animal   both

Name _____ Date _____

Represent What is one thing that only plants need to survive? What is one thing that only animals need to survive?

What is one thing that both plants and animals need to survive? Draw or write your answers in the chart.

Plants	Both Plants and Animals	Animals

Connections in Science

Systems and System Models How do plants and animals work together to survive? Write about an example.

_____

_____

_____

_____

Name _____ Date _____

Complete the "I can..." statement.

I can <u>use models</u>

I can <u>observe</u>
<u>patterns</u>

I can <u>know things in</u>
<u>nature</u>

Name _____ Date _____

Choose an animal and draw what it eats. Tell if your animal is an herbivore, carnivore, or omnivore. Draw and write your answers below.

Name _____ Date _____

1. Do all plants need the same amounts of
the same things?

yes   no

2. **Justify** Draw an example of something
plants need below. Do all plants need
the same amount? Draw or write what
you think.

Name _____ Date _____

3. Do all animals need the same amounts of the same things?

yes   no

4. **Justify** Draw an example of something animals need below. Do all animals need the same amount? Draw or write what you think.

Name _____ Date _____

Essential Question

What do plants and animals need to live?

_____
- -
_____
- -
_____
- -
_____
- -
_____
- -

Word Box

living	nonliving	need	air
water	light	soil	nutrients
survive	shelter		

Name _____ Date _____

What is a plant or animal that you would like to care for? What are the things your plant or animal would need? Draw or write your answers.

_____

- -

_____

- -

_____

- -

_____

- -

_____

Science in Action

Rate Yourself. Circle the image that tells how well you did.

I did use models.

Almost Always   Sometimes   Still Learning

I did observe patterns.

Almost Always   Sometimes   Still Learning

I did know things in nature work together.

Almost Always   Sometimes   Still Learning

Name _____ Date _____

Science in My World
Places Plants Grow

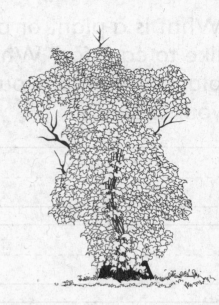

What do you wonder about where plants grow? Write a question.

_____

- -

_____

- -

_____

- -

_____

- -

_____

Essential Question

Where do different kinds of plants grow?

Kayla
Landscape Architect

Name _____ Date _____

Where do plants grow? Draw your observations.

Science in Action

I will use models.
I will observe patterns.
I will know things in nature
work together.

Name _____ Date _____

Inquiry: Where Plants Grow

Go outside with your class and look for two places that plants grow. Where did you see two different plants growing? Draw or write your answers in the chart.

Places Plants Grow
1.
2.

Connections in Science

Systems and System Models Why are these good places for plants to grow? Write what you think.

_____

- -

_____

- -

_____

Name _____ Date _____

Think about one place you found a plant growing. What details did you observe about where the plant lives? Draw or write your observations.

Location 1

Name _____ Date _____

Think about another place you found a plant growing. What details did you observe about where the plant lives? Draw or write your observations.

Location 2

Name _____ Date _____

Complete the "I can..." statements.

I can <u>use models</u>

I can <u>observe patterns</u>

I can <u>know things in nature</u>

Name _____ Date _____

Clarify Think of two plants that grow in different climates. What is one thing about where the plants grow that is the same? What is one thing about where the plants grow that is different? Draw or write what was the same in the middle. Draw or write what was different in the outside part of the circles.

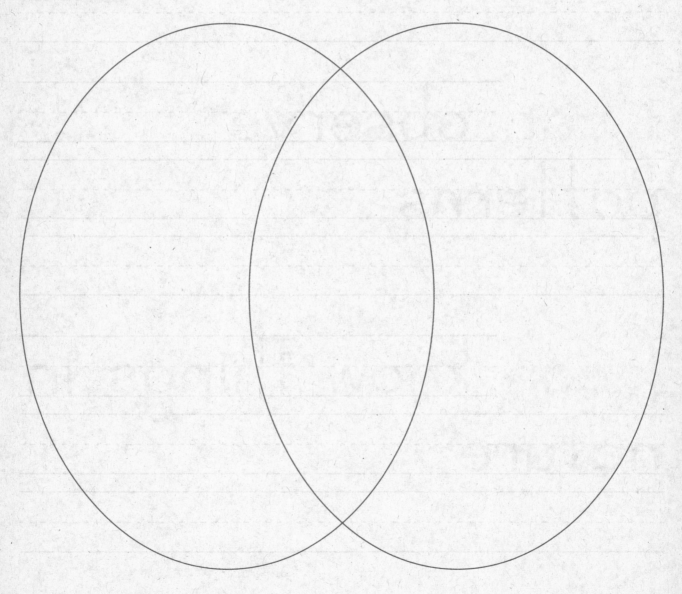

Name _____ Date _____

Interpret Use the chart to tell about plants.
Work with a partner to write or draw two
more answers.

Cause	Effect
1. Grass needs water. We use a sprinkler to water the grass.	1. Grass grows in the yard.
2.	2.
3.	3.

Name _____ Date _____

Essential Question
Where do different kinds of plants grow?

_____
- -
_____
- -
_____
- -
_____
- -
_____
- -
_____
- -
_____

Word Box

forest   desert   arctic   pond

Name _____ Date _____

Why do some plants grow well
in some places and not others?
Explain why different plants grow
in different locations.

_____

- -

_____

- -

_____

- -

_____

Science in Action

Rate Yourself. Circle the image
that tells how well you did.

I did use models.

Almost Always   Sometimes   Still Learning

I did observe
patterns.

Almost Always   Sometimes   Still Learning

I did know things in
nature work together.

Almost Always   Sometimes   Still Learning

Name _____ Date _____

Science in My World
Animal Habitats

What do you wonder about animal habitats? Write a question.

_____

- - - - - - - - - - - - - - - -

- - - - - - - - - - - - - - - -

- - - - - - - - - - - - - - - -

_____

- - - - - - - - - - - - - - - -

_____

- - - - - - - - - - - - - - - -

_____

Essential Question

Where do different kinds of animals live?

Jordan
Animal Trainer

Name _____ Date _____

How can animals live in the arctic?
Draw what you think.

Science in Action

I will use models.
I will observe patterns.
I will know how things in
nature work together.

Name _____ Date _____

Inquiry: Animal Homes

Go on a nature walk with your class. Look for places where animals live. Draw or write what you observe.

Connections in Science

Patterns What are some things that all kinds of animals need to survive?

_____

- -

_____

- -

_____

Name _____ Date _____

Complete the "I can..." statements.

I can <u>use models</u>

I can <u>observe</u>
<u>patterns</u>

I can <u>know things in</u>
<u>nature</u>

Name _____ Date _____

Make a Model: Things Humans Need

Think about where you live. What things do you need every day to survive? Work with a partner. Draw the things that humans need to survive.

Name _____ Date _____

Interpret Choose two of the things that humans need to survive. How do humans get those things? Draw or write what you think.

Name _____ Date _____

1. Choose a bird. Learn the foods your bird eats. Draw and label the different types of food your bird eats.

Name _____ Date _____

Make a Model

2. Draw a home for your bird.
   Label the things your bird needs.

Name _____ Date _____

? **Essential Question**

Where do different kinds of animals live?

_____

- -

_____

- -

_____

- -

_____

- -

_____

- -

_____

- -

Word Box

habitat   ecosystem

Name _____ Date _____

What do the animals need to get from their environment to live? Explain what the animals need using information from your observations and research.

_____

- -

_____

- -

_____

- -

_____

Science in Action

Rate Yourself. Circle the image that tells how well you did.

I did use models.

 Almost Always   Sometimes   Still Learning

I did observe patterns.

 Almost Always   Sometimes   Still Learning

I did know things in nature work together.

 Almost Always   Sometimes   Still Learning

Name _____ Date _____

_____

- - - - - - - - - - - - - - - - -

_____

- - - - - - - - - - - - - - - - -

_____

- - - - - - - - - - - - - - - - -

_____

- - - - - - - - - - - - - - - - -

_____

- - - - - - - - - - - - - - - - -

_____

- - - - - - - - - - - - - - - - -

_____

- - - - - - - - - - - - - - - - -

It's a Wrap!
Plants and Animals

Open Inquiry in Science

1. Rita observes how different plants and animals in an ecosystem depend on each other. Choose an ecosystem and research the different plants and animals that live there. Learn how the plants and animals depend on each other to survive. Make a diorama to model your ecosystem. Use arrows and labels to show how the different living things depend on each other.

2. Ryan observes that people in his community interact with many other living things every day. Discuss how you interact with other living things every day. Record your observations. What other living things do people depend on to survive? Draw or write how people use plants and animals in their environment to live. Share your drawing with your classmates.

Plants and animals depend on each other and other natural resources to survive!

Science in My World
Impacts on Earth's Systems

The woodpecker hits the tree.
What do you wonder about this image?

_____

- -

_____

- -

_____

- -

_____

- -

_____

Look and listen for these words as you learn
about changes in the environment.

Words in Science

environment   burrow   dam   garden

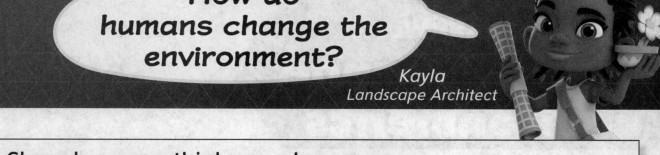

How do humans change the environment?

Kayla
Landscape Architect

Show how you think you change your environment. Draw a picture below.

Science in Action

I will explain my answer.
I will use what I know.
I will know how things in
nature work together.

Name _____ Date _____

Science in My World
Plants Change Environments

The tree changed the sidewalk. What do you wonder about this image?

_____

- -

_____

- -

_____

- -

_____

- -

_____

Essential Question

How do plants change the environment?

Kayla
Landscape Architect

Name _____ Date _____

How do plants change the environment?
Draw or write what you think.

Science in Action

I will explain my answer.
I will use what I know.
I will know things in nature
work together.

Name _____ Date _____

Inquiry: Tree in a Rock

Draw or write what you think will happen in your experiment. Then, draw or write what actually happened.

Talk About It

What will happen to the tree? What will happen to the rock?

Name _____ Date _____

Inquiry: How Plants Change the Environment

Think about what you have learned so far. Draw two ways that plants change their environment. Label them.

Talk About It

How do plants change their environment?

Name _____ Date _____

Complete the "I can..." statement.

I can _explain my answer_

I can _use what I know_

I can _know things in nature work together_

Name _____ Date _____

Before and After Plants Change Environments

Justify Think about the pictures you saw. Think about how one of the plants changed its environment. Think about why that plant changed its environment. Write one sentence about how plants can change their environment.

_____

- -

_____

_____

- -

_____

_____

- -

_____

Talk About It

Why do plants change their environment?

Lesson 1 Plants Change Environments   155

Name _____ Date _____

Plants Changing Environment

Draw two pictures. Draw a before picture of a plant in its environment.

Before

Name _____ Date _____

Plants Changing Environment

Draw an after picture after the plant has changed its environment.

After

Name _____ Date _____

Essential Question

How do plants change the environment?

_____

_____

_____

_____

_____

_____

_____

_____

_____

_____

_____

_____

_____

_____

_____

_____

Word Box

environment

Name _____ Date _____

Think about the tree and the sidewalk.

Are there other ways plants change their environments? How do they change their environments? Draw or write your answer. Use the word from the word box.

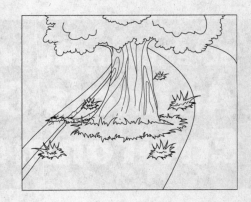

Science in Action

Rate Yourself. Circle the image that tells how well you did.

I did explain my answer.

    Almost Always   Sometimes   Still Learning

I did use what I know

    Almost Always   Sometimes   Still Learning

I did know things in nature work together.

    Almost Always   Sometimes   Still Learning

Name _____ Date _____

Science in My World
Animals Change Environments

What do you wonder about the deer and its environment?

_____
- -
_____
- -
_____
- -
_____
- -
_____

Essential Question

How do animals change the environment?

Kayla
Landscape Architect

Name _____ Date _____

How do animals change the environment?
Draw or write what you think.

Science in Action

I will explain my answer.
I will use what I know.
I will know things in nature
work together.

Name _____ Date _____

Inquiry: Moles Change Their Environments

Think about the mole in the activity. Make a graph from what happened in the activity. Draw how many tufts of grass were left when the mole ate all the worms. Do this twice.

Talk About It

How did the mole change its environment?

Name _____ Date _____

Birds and Seed

Justify Look at the bird and the bird seed.
What might grow where the seed falls?
Draw what you think will happen.

Talk About It

How is the bird changing the environment?

Name _____ Date _____

Complete the "I can..." statements.

I can <u>explain my answer</u>

I can <u>use what I know</u>

I can <u>know things in nature work together</u>

Name _____ Date _____

Squirrels and Nuts

Justify Look at the squirrel. What is it doing?
Draw what the environment will look like if it
buries an acorn.

Talk About It

How is the squirrel changing the environment?

Name _____ Date _____

Beaver Dam

Draw a model of what your river and dam looked like during your investigation. Label your drawing.

Name _____ Date _____

Beaver Dam

Why is the dam a good home for beavers?
Why is it not a good home for other animals?
List two reasons.

_____

The dam is a good home for _____

_____

because _____

_____

_____

_____

_____

The dam is not a good home for _____

_____

because _____

_____

_____

Name _____ Date _____

Essential Question

How do animals change the environment?

_____

- -

_____

- -

_____

- -

_____

- -

_____

- -

_____

- -

_____

- -

_____

Word Box

burrow   dam

Name _____ Date _____

Think about the deer in its environment. What other animals live in the forest with deer?

How do these animals change the forest? Draw or write your answer.

Science in Action

Rate Yourself. Circle the image that tells how well you did.

I did explain my answer.

Almost Always   Sometimes   Still Learning

I did use what I know.

Almost Always   Sometimes   Still Learning

I did know things in nature work together.

Almost Always   Sometimes   Still Learning

Lesson 2 Animals Change Environments   **169**

Name _____ Date _____

Science in My World
People Change Environments

This is a large farming field. What do you wonder about the image?

_____

_____

_____

_____

_____

_____

_____

_____

Essential Question

How do people change the environment?

Kayla
Landscape Architect

Name _____ Date _____

How do you think people change the environment? Draw or write what you think.

Science in Action

I will explain my answer.
I will use what I know.
I will know things in nature work together.

Name _____ Date _____

Inquiry: Plan a Class Garden

Think about your class garden. Draw the plan for your class garden.

Name _____ Date _____

How are humans' needs met?

Think about what humans need. Think about how those needs are met. Draw how two needs are met. Label your drawing.

Name _____ Date _____

Complete the "I can..." statements.

I can <u>explain my</u>
<u>answer</u>

I can <u>use what I</u>
<u>know</u>

I can <u>know things</u>
<u>in nature work</u>
<u>together</u>

Name _____ Date _____

How have people changed the environment?

Interpret Think about your environment. How has it been changed by humans? Make a list of three ways. Draw and label them.

1.

2.

3.

Name _____ Date _____

How I Impact My Environment

Interpret Think about your environment.
How do you change it? Draw one positive way.
Draw one negative way. Label your drawings.

Name _____ Date _____

Plants, Animals, and Humans Impact Environments

Think about how plants, animals, and humans change the environment. Draw three examples of how each has changed the environment.

Plants	Animals	Humans

Name _____ Date _____

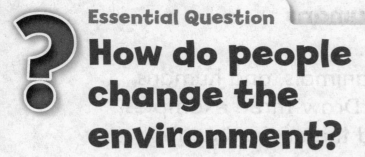

Essential Question

How do people change the environment?

_____

- -

_____

- -

_____

_____

- -

_____

_____

- -

_____

_____

- -

_____

Word Box

garden

Name _____ Date _____

Think about farming.

How does farming change the environment? Draw or write your answer.

_____

- -

_____

_____

- -

_____

_____

- -

_____

Science in Action

Rate Yourself. Circle the image that tells how well you did.

I did explain my answer.

 Almost Always   Sometimes   Still Learning

I did use what I know.

Almost Always   Sometimes   Still Learning

I did know things in nature work together.

 Almost Always   Sometimes   Still Learning

Talk About It

How can you positively change your environment?

_____

- -

_____

- -

_____

- -

_____

- -

_____

- -

_____

- -

_____

- -

_____

- -

_____

- -

_____

It's a Wrap!
Impacts on Earth's Systems

Name _____ Date _____

Let's Make a Model

Design a new city to live in. Your city must include things that meet your needs. How will your city change the environment? Draw a model of your city. Label the drawing.

Plants, animals, and people change environments!

Science in My World
Protecting Our Earth

What do you wonder about
Earth's land, air, and water?

_____

- -

_____

_____

- -

_____

_____

- -

_____

_____

- -

_____

Look and listen for these words as you
learn about protecting our Earth.

Words in Science

pollution	compost	natural resource
conserve	recycle	reduce
reuse		

How can people affect the Earth?

Kayla
Landscape Architect

Show how people can affect the land, air, and water. Draw and write what you think.

Science in Action

I will make observations.
I will see cause and effect.
I will communicate solutions.

Name _____ Date _____

Science in My World
Land, Air, and Water Pollution

What do you wonder about pollution?

_____

- - - - - - - - - - - - - - - - - - - -

_____

- - - - - - - - - - - - - - - - - - - -

_____

- - - - - - - - - - - - - - - - - - - -

_____

- - - - - - - - - - - - - - - - - - - -

_____

- - - - - - - - - - - - - - - - - - - -

_____

Essential Question

 **How do people's actions impact land, air, and water?**

Kayla

184   Lesson 1 Land, Air, and Water Pollution

Name _____ Date _____

Kayla saw people leaving trash on the ground. What can happen if we do this? Draw your answer.

Science in Action

I will make observations.
I will see cause and effect.
I will communicate solutions.

Name _____ Date _____

Inquiry: Land, Air, and Water
Why are land, air, and water important?
Write what you think.

_____

Land is important because _____.

_____

Air is important because _____.

_____

Water is important because _____.

Connections in Science

Cause and Effect What happens when
people pollute the land, air, and water?

_____

_____

_____

Name _____ Date _____

How Humans Impact Land, Water, and Air

Express Think about how humans change or impact the land, water, and air. Identify the causes and effects. Draw or write your examples in the charts.

LAND	
Human Action	**Impact on Earth**

Name _____ Date _____

WATER	
Human Action	**Impact on Earth**

AIR	
Human Action	**Impact on Earth**

Name _____ Date _____

Complete the "I can…" statement.

I can <u>make</u> <u>observations</u>

I can <u>see cause</u> <u>and effect</u>

I can <u>communicate</u> <u>solutions</u>

Name _____ Date _____

1. How can land, air, and water become polluted? Draw or write what you think.

2. **Clarify** How can plants and animals be affected by pollution?

_____

- -

_____

_____

- -

_____

_____

- -

_____

Name _____ Date _____

3. A mascot can be used to represent an action you want people to do. Draw a mascot that will remind people not to pollute. Name your mascot.

Name _____ Date _____

How do people's actions impact land, air, and water?

_____

- -

_____

- -

_____

- -

_____

- -

_____

- -

_____

- -

_____

Words Box

pollution   compost

Name _____ Date _____

How do people affect the land, air, and water in your neighborhood? Draw or write your answer. Use the words from the word box.

Science in Action

Rate Yourself. Circle the image that tells how well you did.

I did make observations.

Almost Always   Sometimes   Still Learning

I did see cause and effect.

Almost Always   Sometimes   Still Learning

I did communicate solutions.

Almost Always   Sometimes   Still Learning

Name _____ Date _____

Science in My World
Help Save Natural Resources

What do you wonder about natural resources?

_____
- -
_____
_____
- -
_____
_____
- -
_____

Essential Question

How can we save natural resources?

Kayla
Landscape Architect

Name _____ Date _____

How can we take care of Earth?
Draw what you think.

Science in Action

I will make observations.
I will see cause and effect.
I will communicate solutions.

Name _____ Date _____

Inquiry: Conserve Water

What happens when you leave the water running while washing dishes? What happens when you turn the water off while washing dishes? Write or draw your observations in the chart.

Water Running	Water Off

Talk About It

Is it better to leave the water running while you wash dishes, or turn it off? Why?

Name _____ Date _____

No More Water

Draw what you think would happen if there
was no water left on Earth.

Name _____ Date _____

How I Use Resources Every Day

Draw a picture to show how you use land, water, and air every day.

How I Use Resources Every Day		
Land	**Air**	**Water**

Put a star next to good choices you can make.
Put an X next to bad choices.

☐ Turn off the water when you brush your teeth.

☐ Water plants outside when they do not need water.

☐ Throw trash in the water.

☐ Only run the dishwasher if it is full.

Name _____ Date _____

Complete the "I can..." statements.

I can <u>make</u>
<u>observations</u>

I can <u>see cause</u>
<u>and effect</u>

I can <u>communicate</u>
<u>solutions</u>

Name _____ Date _____

Count the Resources

Look around your classroom. How many items are made from plants? How many items are made from animals? Put a tally mark in the chart for each item.

Classroom Resources	
Animal Source	**Plant Source**

1. **Interpret** Write one item that was made from a plant source.

_____

_____

2. **Interpret** Write one item that was made from an animal source.

_____

_____

Name _____ Date _____

Help the Environment

Draw a picture to show ways that you can help the environment. Label the ways.

Name _____ Date _____

Essential Question

How can we save natural resources?

_____

- -

_____

_____

- -

_____

_____

- -

_____

_____

- -

_____

_____

- -

_____

Word Box

natural resource   conserve

Name _____ Date _____

How can you take care of our natural
resources? Draw or write your answer.
Use the words from the word box.

_____

- -

_____

- -

_____

_____

- -

_____

_____

- -

_____

Science in Action

Rate Yourself. Circle the image
that tells how well you did.

I did make
observations.

Almost   Sometimes   Still
Always   Learning

I did see cause
and effect.

Almost   Sometimes   Still
Always   Learning

I did communicate
solutions.

Almost   Sometimes   Still
Always   Learning

Lesson 2 Help Save Natural Resources   **203**

Name _____ Date _____

Science in My World
Reduce, Reuse, Recycle

What do you wonder about recycling?

_____

- -

_____

_____

- -

_____

_____

- -

_____

_____

- -

_____

Essential Question
How can we take care of Earth?

Kayla
Landscape Architect

Name _____ Date _____

How can you reduce, reuse, and recycle materials? Draw what you think.

Science in Action

I will make observations.
I will see cause and effect.
I will communicate solutions.

Name _____ Date _____

Inquiry: Sorting Recyclables

What objects can be recycled?
Which objects cannot be recycled?
Sort the objects. Write what properties
you used to sort the objects.

_____

- -

_____

_____

- -

_____

_____

- -

_____

As a class, keep a count of how many plastic
items you recycle during the week. Keep a
count of how many glass items you recycle
during the week. Record the data in the
table below.

Plastic	
Glass	

Name _____ Date _____

Complete the "I can…" statements.

I can <u>make</u>
<u>observations</u>

I can <u>see cause</u>
<u>and effect</u>

I can <u>communicate</u>
<u>solutions</u>

Name _____ Date _____

Compost Design

Represent Design a way to compost materials.
Draw your design. Label the parts.

Name ______________________ Date __________

Reduce, Reuse, and Recycle

Reduce means to use less of something.
Draw things that people can use less of.
Write to explain your drawings.

Name _____ Date _____

Reduce, Reuse, and Recycle

Reuse means to use something again.
Draw things that can be reused.
Write to explain your drawings.

_____

- -

_____

- -

_____

Name _____ Date _____

Reduce, Reuse, and Recycle

Recycle means to make something new from something old. Draw things that can be recycled. Write to explain your drawings.

Name _____ Date _____

Essential Question

How can we take care of Earth?

_____
- -
_____
- -
_____
- -
_____
- -
_____
- -
_____

Word Box

reduce   reuse   recycle

Name _____ Date _____

Look at the picture. What objects can you reduce, reuse, or recycle? Draw or write what you think. Use the words from the word box.

Science in Action

I did make observations.

<table>
<tr><td>Almost Always</td><td>Sometimes</td><td>Still Learning</td></tr>
</table>

I did see cause and effect.

<table>
<tr><td>Almost Always</td><td>Sometimes</td><td>Still Learning</td></tr>
</table>

I did communicate solutions.

<table>
<tr><td>Almost Always</td><td>Sometimes</td><td>Still Learning</td></tr>
</table>

Rate Yourself. Circle the image that tells how well you did.

Talk About It

How can people affect the Earth?

_____

- -
_____

- -
_____

- -
_____

- -
_____

- -
_____

- -
_____

It's a Wrap!
Protecting Our Earth

Open Inquiry in Science

Research how people impact the environment and other living things. Create a solution that shows how people use less natural resources.

1. Josh observes how his class uses natural resources every day. Choose one natural resource that your class uses. Create a plan for your class to use less of the resource during the school year. Draw or write your solution. Share your solution with your class.

2. Kaia observes that people impact the land, air, water, and living things by using natural resources. Record the amount of natural resources your family and friends use. Draw or write how your family and friends can use less natural resources. Explain your solution to your class.

People can protect the earth by reducing, reusing, and recycling!

What are VKVs and who needs them?

" VKVs are flashcards that animate words by kinesthetically focusing on their structure, use, and meaning. VKVs are beneficial not only to students learning the specialized vocabulary of a content area, but also to students learning the vocabulary of a second language. "

Dinah Zike | Educational Consultant
Dinah-Might Activities, Inc. – San Antonio, Texas

Why did you invent VKVs?

" Twenty years ago, I began designing flashcards that would accomplish the same thing with academic vocabulary and cognates that Foldables® do with general information, concepts, and ideas—make them a visual, kinesthetic, and memorable experience. "

I had three goals in mind:

- **Making two-dimensional flashcards three-dimensional**

- **Designing flashcards that allow one or more parts of a word or phrase to be manipulated and changed to form numerous terms based upon a commonality**

- **Using one sheet or strip of paper to make purposefully shaped flashcards that were neither glued nor stapled, but could be folded to the same height, making them easy to stack and store**

Dinah Zike's Visual Kinesthetic Vocabulary®

Why are VKVs important in today's classroom?

At the beginning of this century, research and reports indicated the importance of vocabulary to overall academic achievement. This research resulted in a more comprehensive teaching of academic vocabulary and a focus on the use of cognates to help students learn a second language. Teachers know the importance of using a variety of strategies to teach vocabulary to a diverse population of students. VKVs function as one of those strategies.

An Interview with

Dinah Zike Explaining
Visual Kinesthetic Vocabulary®, or VKVs®

Dinah Zike's
VKV
**Visual
Kinesthetic
Vocabulary**®

How are VKVs used to teach content vocabulary?

"As an example, let's look at content terms based upon the combining form *–vore*. Within a unit of study, students might use a VKV to kinesthetically and visually interact with the terms *herbivore*, *carnivore*, and *omnivore*. Students note that *–vore* is common to all three words and it means "one that eats" meat, plants, or both depending on the root word that precedes it on the VKV. When the term *insectivore* is introduced in a classroom discussion, students have a foundation for understanding the term based upon their VKV experiences. And hopefully, if students encounter the term *frugivore* at some point in their future, they will still relate the *–vore* to diet, and possibly use the context of the word's use to determine it relates to a diet of fruit."

What organization and usage hints would you give teachers using VKVs?

"Cut off the flap of a 6" x 9" envelope and slightly widen the envelope's opening by cutting away a shallow V or half circle on one side only. Glue the non-cut side of the envelope into the front or back of student workbooks or journals. VKVs can be stored in the pocket.

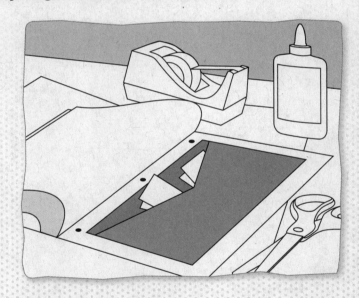

Encourage students to individualize their flashcards by writing notes, sketching diagrams, recording examples, forming plurals (radius: radii or radiuses), and noting when the math terms presented are homophones (sine/sign) or contain root words or combining forms (kilo-, milli-, tri-).

As students make and use the flashcards included in this text, they will learn how to design their own VKVs. Provide time for students to design, create, and share their flashcards with classmates.

Dinah Zike's book Foldables, Notebook Foldables, & VKVs for Spelling and Vocabulary 4th-12th won a Teachers' Choice Award in 2011 for "instructional value, ease of use, quality, and innovation"; it has become a popular methods resource for teaching and learning vocabulary.

✂ cut on all dashed lines   fold on all solid lines

Distance is the amount of space between two places or things.

A **pull** is a force that moves something closer to you.

The students gave the rope a pull.

distance

A **push** is a force that moves something away from you.

The girls push the door open.

Dinah Zike's
Visual
Kinesthetic
Vocabulary®

✂ cut on all dashed lines   fold on all solid lines

t

pull

push

Memory Maker: Draw a story using the words on this VKV.

Memory Maker: Draw yourself at a large distance from school.

Dinah Zike's
Visual
Kinesthetic
Vocabulary®

V K V

✂ cut on all dashed lines   fold on all solid lines

The **planet** we live on is called Earth.

The **Earth** is the third planet from the Sun.

When an object blocks light from the **Sun** it makes **shade**.

Shade is the dark area caused when light is blocked.

The **Sun** is the star closest to **Earth**.

Earth

Dinah Zike's
Visual
Kinesthetic
Vocabulary®

Energy and the Sun

✂ cut on all dashed lines

⬜ fold on all solid lines

Memory Maker: Draw a picture showing the meaning of the words on this VKV.

shade

Sun

planet

Memory Maker: Draw a picture of Earth.

VKV Dinah Zike's Visual Kinesthetic Vocabulary®

✂ cut on all dashed lines

📁 fold on all solid lines

It is a rainy day when the rain falls and makes things wet.

rain

If the sky is cloudy, it has many clouds.

When the wind blows hard, the weather is called windy.

VKV

Dinah Zike's
**Visual
Kinesthetic
Vocabulary**®

Weather

✂ cut on all dashed lines

▭ fold on all solid lines

y

Memory Maker: Draw what it would look like outside if the weather was cloudy, rainy, and windy.

wind

cloud

Dinah Zike's
Visual
Kinesthetic
Vocabulary ®

✂ cut on all dashed lines

📄 fold on all solid lines

fall season

1. Summer is the time of year after spring.

2. It is hottest in summer.

1. Spring is the season after winter.

2. Fall is the season after summer.

1. Winter is the time of year after fall.

2. It is coldest in winter.

Weather

✂ cut on all dashed lines   ⬚ fold on all solid lines

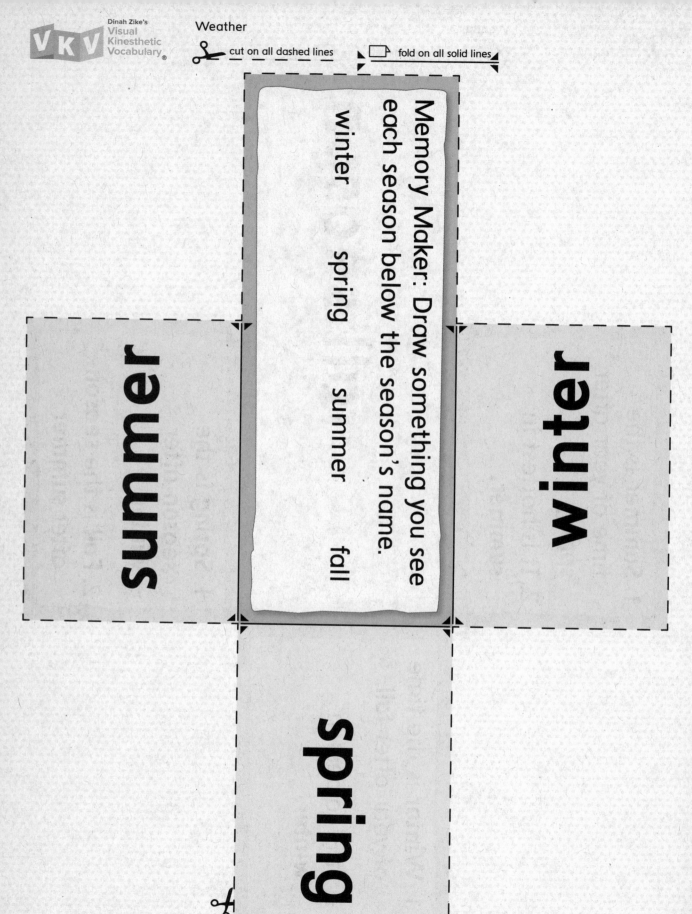

Memory Maker: Draw something you see each season below the season's name.

winter   spring   summer   fall

summer

winter

spring

Dinah Zike's
Visual Kinesthetic Vocabulary®

✂ cut on all dashed lines   ⬚ fold on all solid lines

forecast

Memory Maker: Draw a weather forecast showing weather patterns.

The **weather** is what the sky and air are like each day.

Dinah Zike's
VKV
Visual
Kinesthetic
Vocabulary®

Weather

✂ cut on all dashed lines   ▢ fold on all solid lines

weather patterns

weather patterns

To **forecast** means to say that something will happen by using information about the **weather**.

1. **Patterns** are the repeated way in which something happens.

2. **Weather patterns** are the repeated way the **weather** happens.

VKV

severe weather

1. A sunny day is full of sunlight.
2. Underline the word you see in both *sunny* and *sunlight*.

The **Sun** is the star closest to **Earth**.

Sun

Severe weather means there are very strong conditions outside.

Dinah Zike's
VKV
Visual
Kinesthetic
Vocabulary ®

Memory Maker: Draw what severe weather looks like.

ny

Memory Maker: Draw a sunny day.

The **weather** is what the sky and air are like each day.

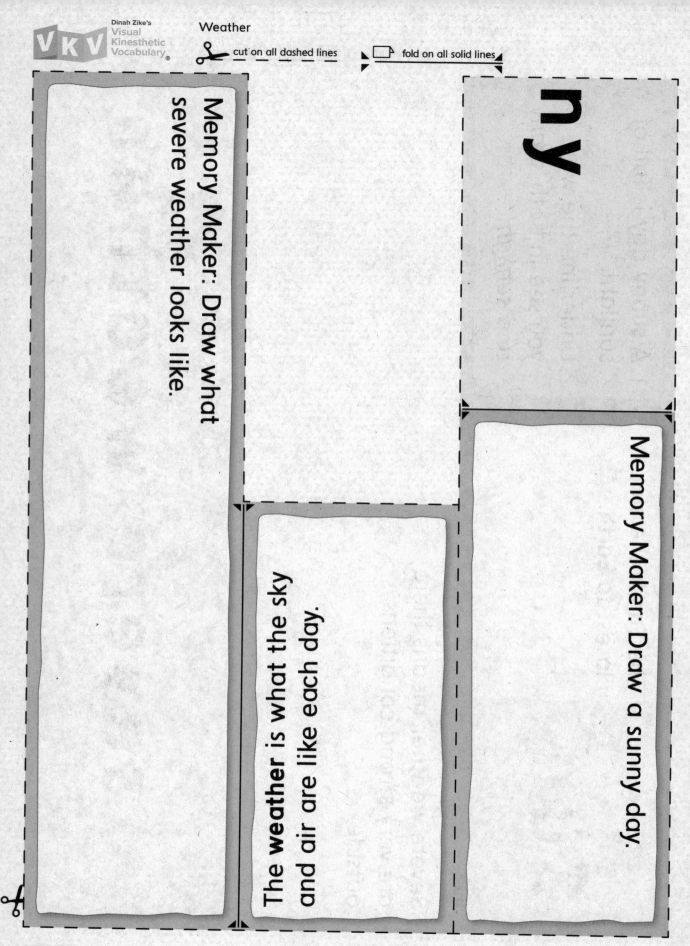

Dinah Zike's
Visual
Kinesthetic
Vocabulary®

A **thunderstorm** is a storm with thunder, lightning, rain, and wind.

A **hurricane** is a strong storm with winds that form a cloud that looks like a funnel.

storm

A **tornado** is a strong storm with heavy rain and winds that blow in a circle.

VKV Dinah Zike's
Visual
Kinesthetic
Vocabulary®

Weather

✂ cut on all dashed lines   ⬜ fold on all solid lines

Memory Maker: How is a tornado different from a hurricane? Draw or write your answer.

hurricane

thunder

tornado

Memory Maker: Draw what happens during a thunderstorm.

Dinah Zike's
VKV
Visual
Kinesthetic
Vocabulary®

✂ cut on all dashed lines

fold on all solid lines

A **nonliving** thing is a thing that does not grow and change, or need food, air, or water to survive.

A **living** thing is something that grows, changes and needs food, air, and water to survive.

Draw an animal.

An **animal** is a living thing that is not a human or a plant.

living

Draw a plant.

A **plant** is a living thing that has leaves, roots, and makes its own food.

VKV — Dinah Zike's Visual Kinesthetic Vocabulary®

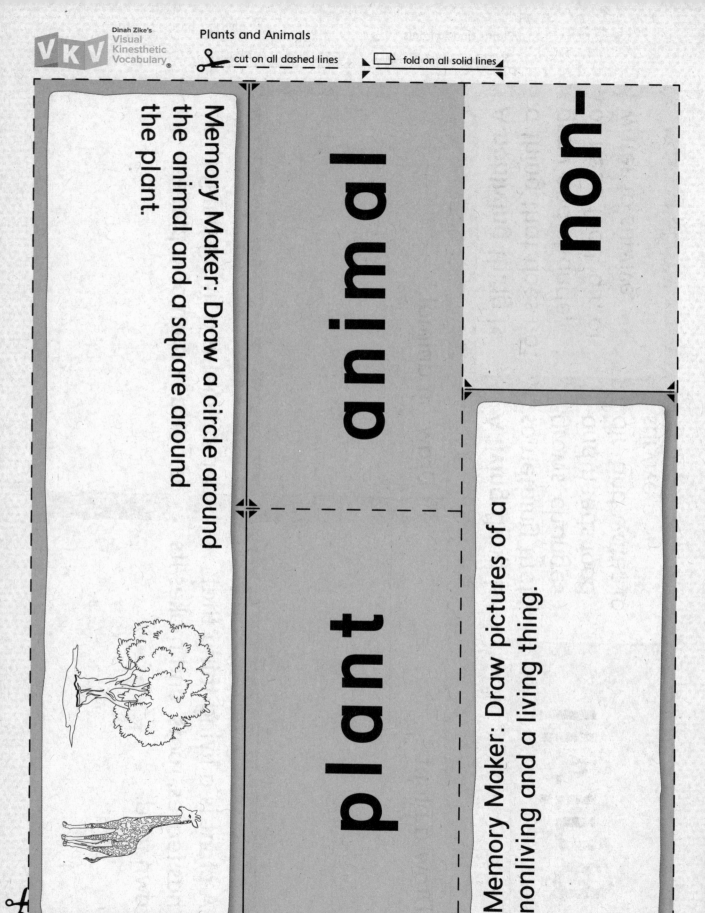

non-

animal

plant

Memory Maker: Draw a circle around the animal and a square around the plant.

Memory Maker: Draw pictures of a nonliving and a living thing.

Dinah Zike's
Visual
Kinesthetic
Vocabulary®

cycle

A life cycle is how a living thing grows, lives, and dies.

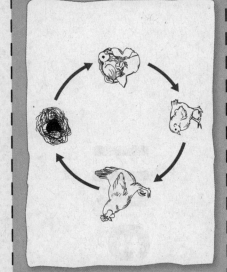

Air is the invisible gas that people and animals breathe.

Soil is the top layer of Earth.

Water is the liquid that falls as rain from the sky.

Dinah Zike's
V K V Visual
Kinesthetic
Vocabulary®

Plants and Animals

✂ cut on all dashed lines ▭ fold on all solid lines

air

soil

water

life

Memory Maker: Draw a picture using all three VKV words.

Memory Maker: Draw a life cycle.

✂ cut on all dashed lines   fold on all solid lines

Circle the animals that live in burrows.

A **burrow** is a hole or tunnel in the ground that an animal makes to live in.

burrow

A **garden** is an area of land used for growing flowers or vegetables.

flower garden

VKV®
Dinah Zike's
Visual
Kinesthetic
Vocabulary®

✂ cut on all dashed lines

📋 fold on all solid lines

Memory Maker: Create your own garden, include as many flowers or vegetables as you like.

ing

Memory Maker: Draw the burrow you want to live in if you are a burrowing animal.

vegetable

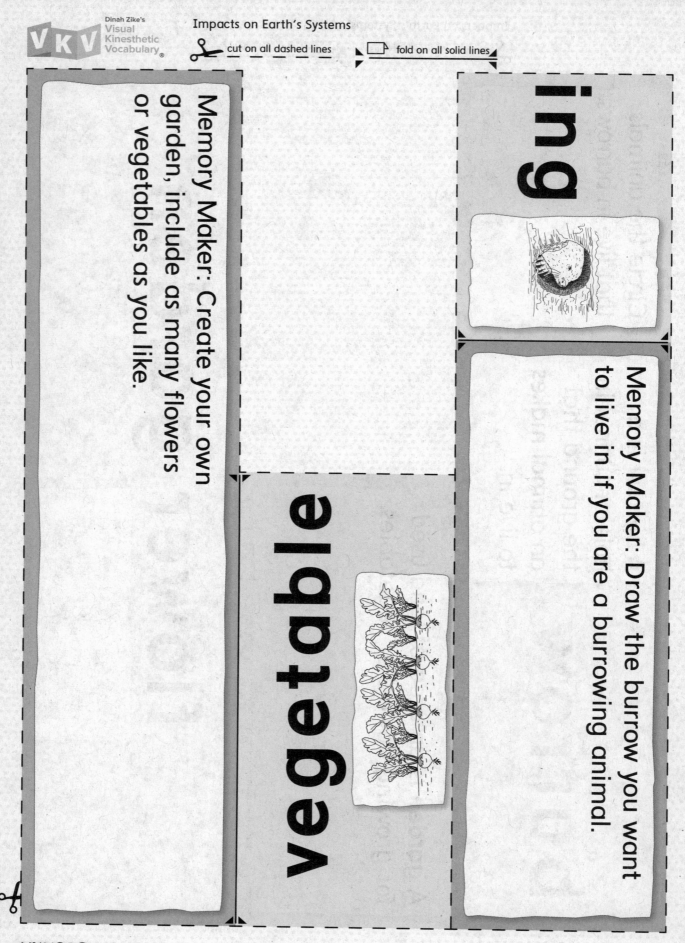

Dinah Zike's
Visual Kinesthetic Vocabulary ®

✂ cut on all dashed lines

◻ fold on all solid lines

Pollution is anything harmful in the air, land, or water.

Compost is a mixture of dead plants.

To **conserve** means to save, keep, or protect something.

pollution

compost

conserve

Dinah Zike's
**Visual
Kinesthetic
Vocabulary** ®

ation

e

e

Memory Maker: Draw a picture of a pile of compost for a garden.

Memory Maker: Draw pollution you have seen in your town.

Memory Maker: Draw a picture that will make people want to conserve trees.

de

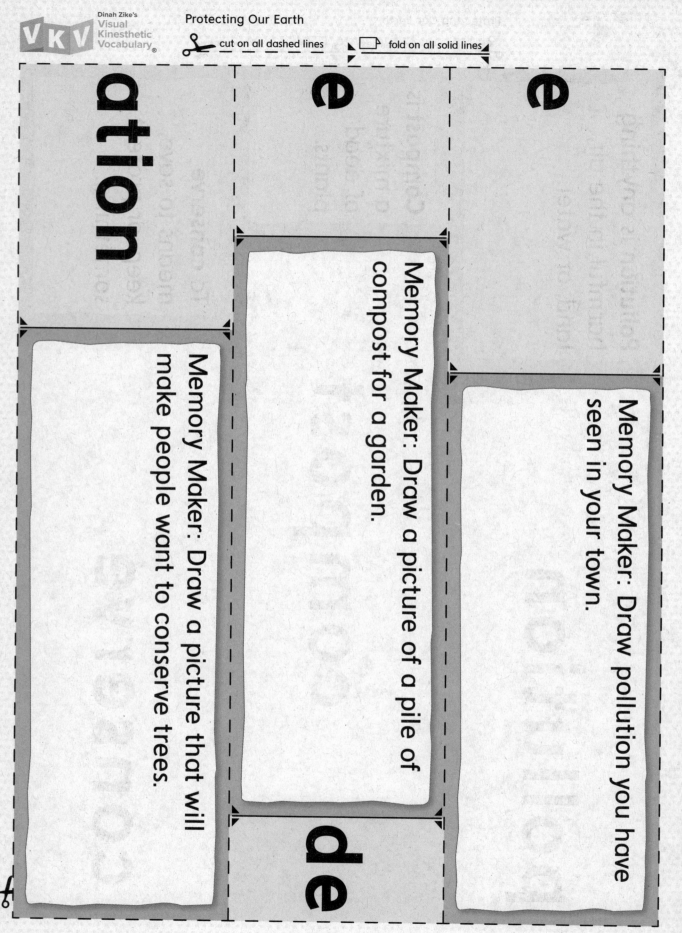

cut on all dashed lines   fold on all solid lines

recycle

1. **Reuse** means to use something again.

2. **Recycle** means to make something new from something old.

Reduce means to use less of something.

VKV

Dinah Zike's
**Visual
Kinesthetic
Vocabulary**®

duce

use

**Memory Maker: Draw a comic strip
showing the words on this VKV.**